QUESTIONS & ANSWERS

GRP
Boat
Construction

Keith Figg and John Hayward

Newnes Technical Books

The Butterworth Group

United Kingdom	**Butterworth & Co (Publishers) Ltd** London: 88 Kingsway, WC2B 6AB
Australia	**Butterworths Pty Ltd** Sydney: 586 Pacific Highway, Chatswood, NSW 2067 Also at Melbourne, Brisbane, Adelaide and Perth
Canada	**Butterworth & Co (Canada) Ltd** Toronto: 2265 Midland Avenue, Scarborough, Ontario M1P 4S1
New Zealand	**Butterworths of New Zealand Ltd** Wellington: T & W Young Building, 77—85 Customhouse Quay, 1, CPO Box 472
South Africa	**Butterworth & Co (South Africa) (Pty) Ltd** Durban: 152—154 Gale Street
USA	**Butterworth (Publishers) Inc** Boston: 19 Cummings Park, Woburn, Mass. 01801

First published 1979 by Newnes Technical Books,
a Butterworth imprint

© Butterworth & Co (Publishers) Ltd, 1979

British Library Cataloguing in Publication Data

Figg, Keith
GRP boat construction. —
(Questions and answers series)
1. Boat-building   2. Fibreglass boats   3. Glass
reinforced plastics
I. Title   II. Hayward, John   III. Series
623.82'07'38   VM331

ISBN 0-408-00317-0

Printed in England by Cox & Wyman Ltd.,
London, Fakenham and Reading

PREFACE

The steady increase in the popularity of pleasure sailing since the early 1960s has been the prime reason for the rapid developments that have taken place in the moulding of glass reinforced plastics boats. It is doubtful whether traditional timber building techniques could ever have satisfied the demand and so, as is often the case, an industrial process has largely replaced a craft. But this is not to say that the production of boats from GRP (or resin-glass, as it is sometimes called) makes no demands on manual skills. On the contrary, and as this little book will amply illustrate, the moulding process itself calls not only for a knowledge of the materials involved and the attendant hazards, but also for a skill that can only come with training and experience.

At the same time, the construction of the plug from which the mould is made involves so many of the techniques of traditional timber construction that the moulder may often decide to complete the plug, once its function in mould making is served, as a sea-going vessel.

Although the enthusiastic boat owner will find much that is of practical value in maintaining and repairing his craft, this book is essentially a no-frills primer for the professional and more ambitious amateur, as well as the great many students and trainees who are about to enter what is now an industry of size and economic importance.

CONTENTS

1

THE GRP HULL

What is GRP?

GRP stands for glass reinforced plastics. It is a composite material consisting of high strength glass fibres bonded together by a resin of lower strength. This results in a material which exhibits the best characteristics of both materials. By varying the quantity of both the resin and the glass, the properties of GRP can be varied to suit a particular design application.

Why is GRP used for boat construction?

It has excellent strength properties, and is strong for its weight and volume. It has a high resistance to water and has good weathering resistance. It has a high resistance to impact and will not shatter. But probably the most important property that has influenced its widespread use is the relative ease with which it can be moulded to fairly complex shapes.

Is there a limit to the size of moulding that can be produced?

Early GRP craft were usually fairly small, but modern moulding techniques enable craft ranging from pulling dinghies to motor yachts, fishing boats and naval vessels of more than 35 m (100 ft) in length to be built. Some modern naval craft exceed 70 m (200 ft).

What proportion of craft are made from GRP?

In the leisure boating industry about 90% of the many types of sailing and power craft are constructed in GRP. Large numbers of workboats are now produced, and the non-magnetic properties of GRP make it an alternative to wood for the construction of minesweepers.

Why are most yachts and other pleasure craft constructed of GRP?

For pleasure craft and sailing yachts up to about 10 m (30 ft) in length and produced in quantity, it is cheaper. GRP boats can be built more quickly than traditional wooden craft and make less demands on skilled labour. Time consuming jobs like caulking and painting (a timber hull requires at least six coats of paint) are eliminated, and only an antifouling and an optional boot-top are required.

What is the main advantage of GRP to the owner or user?

For the user the greatest advantage is the reduction in maintenance. Annual varnishing and painting are unnecessary and other tasks like the periodic caulking of seams are eliminated. However, GRP boats are not maintenance-free, and the owner must adopt a strict maintenance schedule and set aside time for maintenance and repair of minor damage if the boat is to serve a useful and trouble-free working life (see Chapter 8).

The reduction in maintenance can appreciably reduce the running costs of a GRP boat when compared to timber.

Are there any other advantages in GRP?

Yes. As smaller GRP craft, such as sailing yachts and power boats, are constructed from only two major components (the hull and the deck) the number of points at which leaks can develop is drastically reduced. A well built GRP boat is completely watertight, and should remain so for many years. Except in well found craft of the highest standard of construction, a wooden boat will often develop minor leaks after a long period of heavy use.

Does the method of construction itself offer any advantages?

Yes. The GRP hull and deck consist of single homogeneous mouldings and are therefore able to dispense with the internal structure which holds a wooden boat together. Thus frames, heavy floor timbers, stringers and beam shelves are eliminated, as in most cases are deck beams. The structural shell of a GRP boat is usually less than 20 mm (¾in) thick, whereas the total thickness of hull structure in a wooden boat is usually between 40 and 60 mm (1½ and 2½ in) at the frames, and between 60 and 100 mm (2¼ and 4 in) at bilge stringer and beam shelf. This gives considerably more usable space below decks, and makes the GRP boat easier to clean (Fig. 1).

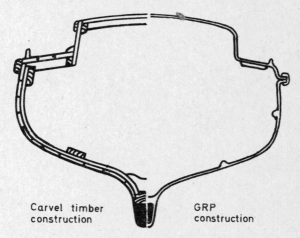

Carvel timber
construction

GRP
construction

Fig. 1. Section through a shoal draft cabin yacht comparing
traditional timber construction with GRP

Unlike timber, GRP is resistant to attack by marine borers, and is resistant to decay.

A further advantage is that the amount of metal in the form of fastenings is much reduced and electrolytic corrosion, one

3

of the major bugbears of wooden construction, is unlikely to occur.

Are there any disadvantages?

Yes. Unless well insulated and ventilated condensation can be a problem in colder temperatures. Also, because it is harder and thinner, a GRP boat will invariably be noisier in operation than one built from wood.

Although maintenance is reduced, painting of the bottom, and of the topsides when this becomes necessary, are more difficult than with a wooden boat and the specially formulated plastic paints are more expensive.

Repair of minor damage is generally no more difficult than with a wooden boat, but more extensive structural damage to hull or deck is costly to repair and must, like the moulding process, be carried out in controlled conditions. In cases of serious damage the hull or deck moulding may need to be returned to the manufacturer for cutting out the damaged section and laying up the repair piece in the mould.

Serious structural damage to a wooden (or steel) boat can usually be repaired by any boatyard with the available skills.

What are the relative merits of GRP and steel?

For vessels of less than 12 m (40 ft) steel is heavier than GRP. Reducing the weight of so small a steel hull would mean reducing the skin thickness to such an extent that it would wrinkle between frames, would easily dent, and would quickly corrode through if the surface coating were broken. Corrosion is a much greater problem in small steel craft than with wood, and contact between the hull and non-ferrous objects (brass screws, coins, and even nickel silver cutlery!) must be avoided.

Noise and condensation are even more troublesome than with GRP, and maintenance is more difficult.

Nevertheless a number of successful smaller steel yachts have been made, mainly in Holland, where the much lower salt content of their inland waterways causes little corrosion.

4

How much skill is needed in making a GRP boat?

GRP boat construction in factory conditions is an industrial process and a thorough knowledge is required of the materials, hazards, and basic procedures. Although experience in traditional construction methods is not essential it can be an asset and special training for the trade often includes woodwork and the construction of traditional hulls. Such knowledge is essential for the construction of the wooden 'plug' from which the moulds for any new design of craft are made.

Skills required of the individual worker are less than those of the traditional boatbuilder, whose job is more varied because of the far greater complexity of the structure of the wooden boat. He is therefore more quickly trained.

However, the same pride in workmanship should remain as dire results attend every attempt to cut corners — in GRP construction as much as in any other.

2

MATERIALS

What resins are used in the manufacture of GRP?

Many resins can be used; the ones most commonly employed
are called unsaturated polyester resins. These are thermo-
setting materials, the derivation of which is shown in Fig. 2. In
appearance they are viscous pale-coloured liquids. Polyester
resins will in the course of time change to a solid or cure by
themselves; however for laminating purposes the curing time
has to be compressed into practical limits by the addition of
certain chemicals which will be explained later.

Many different types of resin are available. Which are the most common and what are their uses?

General purpose resins   These are polyesters formulated for
general moulding work by the hand lay-up method and are
available in a wide variety of viscosities to suit the moulder's
requirements. A wide range of physical and mechanical properties is available.

HET acid resin   This type of resin is commonly used in the
USA. It is similar to the general purpose resins in its properties.
However, its fire resistance is superior to resins made self-
extinguishing by other methods.

Teophthalic resin   These resins have improved mouldability
characteristics and possess superior chemical resistance.

Gel coat resin   This is specially formulated for use as the gel
coat, to give tough resilient films.

Fire retardant resins These contain cholorine compounds in the resin to improve its resistance to flame, or to make it self-extinguishing.

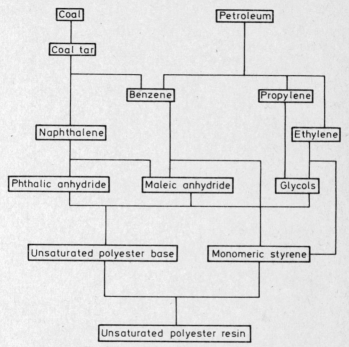

Fig. 2. Manufacture of polyester resins

What is meant by curing of resins?

The resins used in laminating are unsaturated polyester resins. This means that they are capable of being 'cured' from a liquid to a solid when subjected to the right conditions. This is the term commonly given to the polymerisation reaction, which is an irreversible process.

How are resins cured?

The resin consists of a solution of polyester in a monomer, normally styrene. The styrene enables the resin to change from a liquid to a solid by cross linking the chains of the polyester. This is achieved at room temperature by the addition to the resin of a catalyst and accelerator.

What are catalysts and accelerators?

Catalysts are added to the resin to initiate the polymerisation reaction, whilst accelerators are added to enable the reaction to proceed at normal temperatures. All accelerators exert little influence on the resin in the absence of catalyst; they are sometimes added to the resin by the manufacturer and are known as pre-accelerated resins. The various combinations of catalyst and accelerators in use are as follows:

  (a) Methyl ethyl ketone peroxide (MEKP) and a cobalt accelerator.
  (b) Cyclohexanone peroxide and a cobalt accelerator.
  (c) Benzoyl peroxide and a tertiary amine accelerator.

For easier handling, mixing and safety purposes, catalysts and accelerators are normally manufactured as a mixture with another chemical, either in paste or liquid form. The quantities of catalyst and accelerator used in the resin are normally recommended by the manufacturer and range from 1% to 4%. The gel time may be varied by adjusting the accelerator content and not the catalyst. However, if they are both used in too high or too low a concentration, they will act to inhibit the polymerisation reaction.

What are glass fibres?

These consist of continuously-spun glass filaments, usually of a diameter between 8 and 14 μm. These are then collected into bundles of approximately 200 filaments, which form the basic strand.

What type of glass is used?

The type of glass most commonly used is known as 'E' glass, which is a low alkali borosilicate glass with an alkali content of less than 1%.

Why is glass fibre used as a reinforcement?

There are a number of reasons for using glass fibre as a reinforcement material:
- (a) It has good mechanical strength.
- (b) It is very stable under extremes of temperature and humidity.
- (c) It does not deteriorate or perish.

In what form is glass fibre available?

Glass fibre is available in a wide variety of forms and listed below are those more commonly used in the boatbuilding industry.

Chopped strand mat (CSM)   This is a low cost general purpose reinforcement. The mat comprises 1- to 2-in long chopped strands of glass which are bonded or held together in a random manner.
   It is available in several grades depending on the type of glass and binder used. In boat construction mats with a medium solubility binder are generally used, which can be wetted out slowly to allow the mat to retain its structural properties during the initial stages of the lay-up. The mat is available in various weights and is normally specified in grammes per square metre (GSM), though ounces per square foot or per square yard may be found.

Rovings   These consist of bundles of continuous strands and resemble a loose untwisted rope. They are normally specified by the number of strands, 60 being the most common. Rovings are employed as undirectional reinforcements, woven into coarse fabrics, chopped into short lengths for mats or used for spray moulding.

9

Woven rovings These consist of rovings that have been woven into a heavy fabric. The rovings are not spun as in a cloth but woven as strands into a plain square pattern. They are available in a number of different weave patterns and weights, and form a reinforcement which is intermediate in cost and properties between chopped strand mat and woven cloth.

The main advantages are: high glass content, good handling characteristics, high directional strength, and good resistance to impact. However it does possess the following disadvantages: it is difficult to wet out, has poor inlaminar adhesion, the weave tends to trap air bubbles, and it can cause resin richness.

What are sizes, finishes and binders?

A size is a chemical treatment applied during the manufacture of filaments immediately after they are formed. The function of sizes is to protect the filaments against damage by abrasion due to mechanical handling, to hold the filaments together in a strand and to render the glass more readily wetted by the resin to obtain good adhesion. A finish is a chemical treatment applied to a cloth after it is woven and cleaned, the function of which is to improve the chemical bond between the resin and the cloth. A binder is the bonding resin used in the manufacture of chopped strand mat, to hold the fibre bundles together during the laminating process.

What is the gel coat?

This is the outermost layer of the laminate, which provides the surface finish. It forms a barrier which seals off the laminate and prevents penetration by oil and water of the laminate by capillary action. The gel coat resin is normally compounded from the mixing resin and in use is brushed or sprayed on to the mould in an even layer about 0.4–0.5 mm (15–20 thousandths of an inch) thick.

What are fillers?

A filler is an inert substance usually of powder form added to the resin for a number of reasons of which the most important are: to reduce the resin content of the moulding (i.e. cost), to facilitate moulding, and to give specific properties to the resin. The filler should possess the following qualities: it should not inhibit the resin cure, it should not age or deteriorate under service conditions, it should not affect the physical and mechanical properties of the laminate, and it should be inert. Many materials are used, such as chalk or slate powder for bulking of the resin and antimony oxide for self-extinguishing mouldings. Fillers should not be used in hull construction or highly stressed parts. Lloyds rules only allow the use of calcium carbonate.

What are pigments?

Pigments are used to impart a through colour to the resin, to hide fibre pattern, and to reduce the effects of ultra-violet light. The pigments used should not inhibit the resin cure, and should not be affected by organic peroxides or cobalt accelerators, also they should be unaffected by the heat generated (or exotherm) during cure. They are generally limited to 3–5% of the resin weight used.

Can GRP be made fireproof?

GRP cannot be made completely fireproof. However it can be made fire-retardant. This is achieved either by chemical additions to the resins or by coating the resin with a fire-retardant resin. The latter method is commonly used in boatbuilding. The resins used are known as *intumescent* resins. These have the property of swelling up under the action of flame and form a protective carbonaceous coat which acts as a surface insulation.

3

PLUG AND MOULD CONSTRUCTION

What is a mould and what is its function?

The majority of professionally made GRP hulls are moulded, that is to say, the shape of the hull and its outer surface finish are attained by *laying-up* the glass-resin laminates inside a shell which is the exact shape of the required hull (or deck). When the GRP has cured sufficiently the mould, which is generally either in two or three pieces, is parted to release the hull. A mould such as this, of which the inner surface forms the outer shape of the required moulding, is called a *female* mould (Fig. 3).

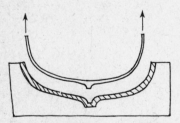

Fig. 3. A simple dinghy hull mould-ing is withdrawn from a female mould

How is the mould produced?

To produce a mould from which a boat hull will be made, a shape or pattern known as a *plug* must first be constructed.

SOME STABLE TIMBERS

Name	Average weight, pounds per cubic foot air dried	Seasoning movement in service	Properties: drying rate	Woodworking resistance in cutting	Properties: blunting effect
Abura	36	Small	Rapid	Medium	Moderate
Agba	30	Small	Rather slow	Medium	Moderate
African mahogany	35	Small	Fairly rapid	Medium	Moderate
Yellow pine	24	Small	Rapid	Low	Mild
Western red cedar	24	Small	Rapid	Low	Mild
Obeche	24	Small	Very rapid	Very low	Mild
Jelutong	28	Small	Rapid	Low	Low

The plug must be precise in shape and size, with a surface finish equal to that of that required for the intended GRP boat hull. The mould is then made around the plug, often from GRP.

How are accuracy and surface finish achieved?

The accuracy and finish of the plug, and therefore of the moulded hull, depend on a number of factors. Of prime importance is that the initial lofting and subsequent mould work is carried out by competent craftsmen. A high degree of workmanship must be maintained throughout the whole of the plug production, as any inaccuracy or defect will be repeated in all the craft that are produced from the mould for which the plug is being constructed.

From what materials is the plug usually made?

Most plugs are made from timber, which must be of a dimensionally stable variety to ensure that the plug does not change shape during or after construction. To maintain stability throughout the building process, timber should be kiln dried with a moisture content of not more than 10–12%. Suitable timbers are listed on p. 13.

How is stability ensured during construction of the plug?

The timber should be stored or kept in the same building in which the plug will be made, and the workshops should be free of draughts and an even temperature maintained. Preferably the plug should be constructed in the shop where the eventual production of the mould will be carried out. To maintain stability and fairness, movement of the plug from one shop to another, which could cause moisture and temperature changes, should be avoided.

The mould laying-up process should be started as soon as possible after completion of the final surface of the plug.

How is the plug constructed?

Strip planking is the most common method used on plugs from 5 to 11 m (15 to 35 ft) long, particularly on deep keeled yachts where it may be too difficult, because of the tuck, to use diagonal planking.

Planks are laid edge to edge on closely spaced frames of not more than 450 mm (18 in) apart. These edges are often spindled to form hollows and rounds so that a close fit is maintained on curved sections without the need to shoot and bevel each plank. Because it may be necessary to take up the shape of the planks at an early stage, an initial shaped garboard can be fitted (Fig. 4).

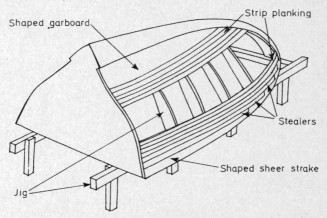

Fig. 4. Use of a shaped garboard and sheer strake enables parallel strips to be used for planking

Timber often shrinks after sawing or planing, even when seasoned. How is this minimised?

The effects of timber shrinkage are minimised by using quarter sawn (rift sawn) timber wherever possible (Fig. 5).

Jelutong is the recommended timber for planking, being

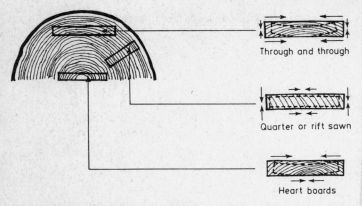

Through and through

Quarter or rift sawn

Heart boards

Fig. 5. Rift or quarter sawn boards are cut on a radius and do not distort or warp in section as do through and through (tangent sawn) boards. The arrows show direction and relative amounts of shrinkage, and the dotted lines show shape after shrinkage

very stable and gluing well. It is easily worked to a fine finish. However, even this timber can move when one side is sealed and one side open to the atmosphere, as is the case with the plug. It is essential therefore that the mould lay-up procedure is started as soon as possible.

How are the planks fixed?

Planks are usually glued and fastened edge to edge with occasional through fastenings into the frames.

Are there any other methods of plug construction?

Yes. Cold moulded construction can be used. This is a process that came to the fore in World War II for the construction of high speed wooden aircraft such as the Mosquito, and was much favoured by the designer Uffa Fox for his wartime work on lifeboats and subsequently for sailing yachts such as the Fairey Atlanta.

The process uses knifed veneers rather than sawn timber, the veneers usually being 75–100 mm (3–4 in) wide and 3–9 mm ($^1/_8$–$^3/_8$ in) thick.

How is a cold moulded plug made?

The veneers or planks are laid on a closely spaced frame or ribbed skeleton at about 45° to the centreline (Fig. 6).

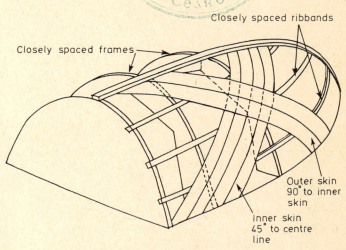

Closely spaced ribbands

Closely spaced frames

Outer skin 90° to inner skin

Inner skin 45° to centre line

Fig. 6. Multi-skin diagonal planking of the plug

Planking normally starts about amidships and away from the stem, with the following layer at right angles to the first (i.e. towards the stem). Subsequent layers are laid 'turn and turn about.'

17

If wooden boats can be built in this way, surely the plug, once a satisfactory mould has been made from it, can be fitted out as a boat?

This is often done, and goes a long way to offset the very high cost of moulding. In such a case the plug would be constructed as if building a boat, with an actual backbone of stem, keel and stern post. The first layer of planks would be stapled to the jig and only the following layers glued. This would enable the finished shell to be lifted off the jig after the moulding sequence and fitted out. If the plug is to be used as a boat after the moulding sequence, all fastenings used in its construction must be non-ferrous, and all adhesives must be of approved types for immersion in water.

It is also possible to build a plug in the conventional carvel manner (see Questions and Answers on Wooden Boat Construction, in this series) and there have been occasions when moulds have been made from the hulls of successful wooden boats.

Can GRP be used for plug construction?

It often is, in the form of C-Flex, particularly where the cost of plug production must be minimised.

How does this method differ from cold moulding?

It is similar. A wooden framework comprising backbone and closely spaced frames is constructed, and to this is attached the C-Flex, which is stretched fore and aft. The C-Flex is held in place by skewers or picks and is then impregnated with a light application of low viscosity resin. Further layers of glass are then added (chopped strand mat or woven roving, see Chapter 2) with resin until the desired thickness is attained.

Quite a large amount of filler and subsequent rubbing down are usually required with this method.

***Are there any special design considerations that apply to plugs
and not to wooden boats?***

Yes. The hull must be designed to facilitate mould release and
this places certain restrictions on the shape and detail design:

1. There must be no undercuts on one-piece moulds.
2. Rubbing strakes, bilge keels and other appendages must
   usually be omitted from the mould and plug. They are,
   of course, fitted to the hull moulding.
3. Angles of bilge, counter, chines, etc. must be such that the
   mould can be easily withdrawn. At least 2−3° declivity
   (draft angle) must be allowed (Fig. 7).
4. Wherever possible all corners should be rounded to a
   radius of at least 6 mm (¼ in).

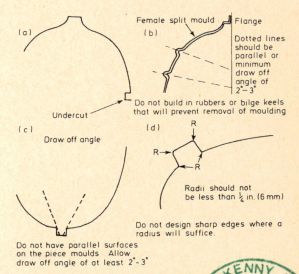

*Fig. 7. Adequate draw-off (draft) angles and generous radii
are essential in good mould design. Avoid undercuts and
parallel surfaces*

19

How is the very high finish required of the mould achieved?

By ensuring a high finish on the plug. There are various ways of doing this but one well-tried and recommended way is:
1. Prime bare wood with a full coating of polyester resin, preferably with the addition of a dark pigment.
2. When fully cured flat down with wet and dry.
3. If any undulations can be seen on the surface apply polyester filler and flat down again.
4. Apply a final coating of furane resin, either by brush or spray.
5. Cut back with wet and dry abrasive paper every four or five coats until a satisfactory finish is attained.
6. Buff to a brilliant shine with increasingly fine degrees of cutting compound. It is often useful at this stage to scribe in the waterline, boot-top, upper finished edge of hull, skin fitting locations, etc. The surface is now ready to receive the release agents prior to the moulding sequence.

From what material is the mould made?

Most moulds are made from GRP, though for limited production runs timber moulds have been used.

The lay-up procedure is similar to that used for the hull (see Chapter 4) with the following differences:
1. To maintain the proper shape the mould thickness is usually 1½ times the hull moulding thickness.
2. To allow for any probable reburnishing of the mould surface the gel coat thickness is increased by approximately 0.6–0.75 mm (0.025–0.030 in).
3. To reduce shrinkage and keep down moulding costs 25% filler may be added to the resin mix.

From how many parts is the complete mould made?

This depends on the shape and size of the vessel to be moulded. The simplest form of mould is the one piece mould, but this is only generally used for small craft such as sailing and pulling dinghies. For larger craft, particularly of more complex shape, the mould is made from two or three pieces, the number being dictated by such features as tumble-home (the sloping in of the hull sides) which is a characteristic of many modern class yachts. Undercuts and reverse bevels also make the one-piece type of mould impracticable. Fig. 8 shows a simple three-piece mould.

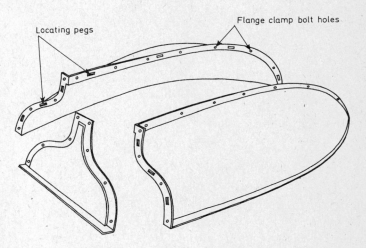

Fig. 8. A three piece dinghy mould showing locating pegs

To form a multi-piece mould, shaped pieces of timber of which the moulding surface has been waxed and prepared are affixed to the finished surface of the plug at an angle of 90° to form a flange. The shape of the plug will then dictate the location of the flanges to facilitate mould release. Small pieces of timber are then attached to the wooden flanges. During the moulding process these will form locating sockets. Having sealed the gaps and corners with a moulding wax the section is then gel coated and the necessary layers of reinforcement laid-up.

21

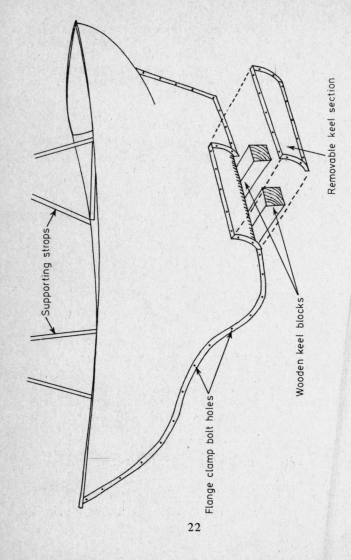

Supporting strops

Removable keel section

Wooden keel blocks

Flange clamp bolt holes

Fig. 9. Method of supporting hull during mould release

22

The wooden flange formers are taken off when the section is cured and the flange surfaces are then waxed and release agent applied.

Finally, the remaining sections are laid up, the mating flange being formed in the process. Holes to accommodate clamp bolts should be bored whilst the mould is in situ.

How important is the fit of the flanges?

The fit of the flange determines the severity of the flash line. The flash line can be ground off and the final buffed surface will show no indication of their existence.

Can any provision be made to facilitate mould removal on larger mouldings?

Smaller sections can be incorporated into the mould to locate bilge blocks and thus support the moulding while the mould is being removed (Fig. 9).

What is a pivot mould?

On large scale production lines the complete mould or mould sections are often made to pivot. This makes the laminator's job easier as the mould can be rotated to present him with a nearly horizontal surface (Fig. 10).

Further advantages are that 'drainage' of the resin is reduced and the mould can be inverted to prevent the gel coat from pickling due to excessive styrene build up.

A small electric motor or hand winch incorporated into the pivot mould framework assists in the movement of large and heavy sections.

Does the finished mould need strengthening in any way?

On small moulds it is usually sufficient to bond a number of bulkhead-like formers to the outer surface (Fig. 11). On larger moulds it is often necessary to bond on longitudinal stiffeners

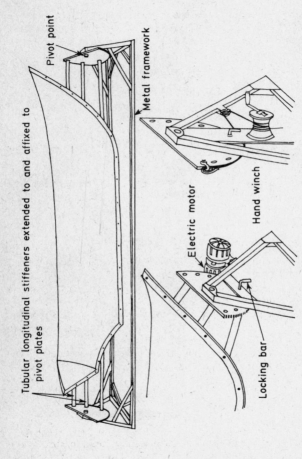

Fig. 10. Simple pivot mould (top), adapted to be turned by motor or hand winch (bottom)

Tubular longitudinal stiffeners extended to and affixed to pivot plates

Pivot point

Metal framework

Electric motor

Hand winch

Locking bar

24

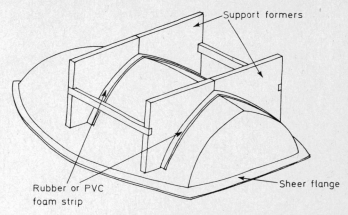

Fig. 11. *Two substantial formers will give adequate support to a simple dinghy mould*

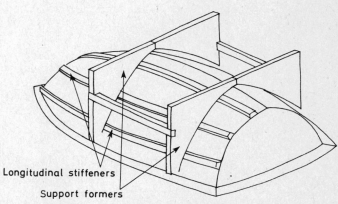

Fig. 12. *Longitudinal stringers attached to the outside of the mould will be necessary for larger mouldings*

25

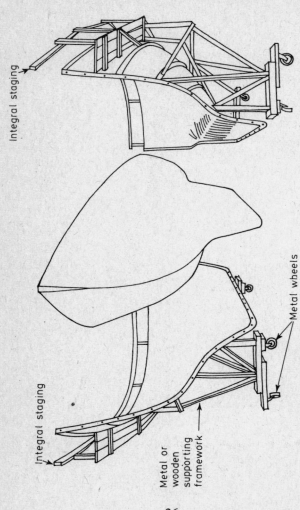

Integral staging

Integral staging

Metal wheels

Metal or
wooden
supporting
framework

Fig. 13. Staging and handrails are incorporated into the two piece mould of this deep keeled yacht

26

(Fig. 12). A layer of rubber or PVC foam is usually placed between the mould and stiffeners before bonding to spread the load and avoid local distortion.

How is access provided for laminating the top of a hull moulding?

Where it is not practicable to use a pivot mould, scaffolding has to be erected or a movable staging provided. However, with careful planning, staging and handrails can be incorporated as part of the mould supports (Fig. 13).

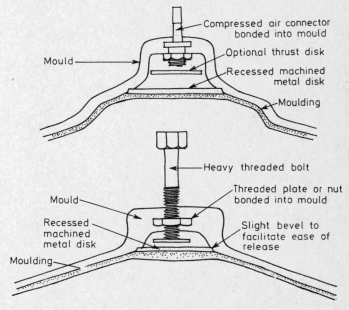

Fig. 14. Bonded in air connector (top) and screwjack (bottom) to facilitate mould release

What other refinements are possible in mould design?

Metal strips can be bonded into flanges to assist the trimming operation. These will also reduce damage to the moulded edge.

Jacking points and pressure valves may also be incorporated in the mould to assist mould release (Fig. 14).

4

FABRICATION

What methods are used to fabricate GRP boats?

Several methods have been successfully used; however the hand lay-up or contact method and spray moulding methods are the two most widely used. Other techniques such as vacuum and pressure bag have been employed in the manufacture of small boats where large production runs offset the large capital costs involved.

What is the hand lay-up (contact moulding) method?

The process is basically the laying up of successive layers of resin impregnated glass reinforcement in a mould of the desired shape, and then allowing the laminate to cure at room temperatures. The advantages of this method are that capital outlay is small and that it is easily used with a wide range of mould shapes and sizes. However, the major disadvantage is that the quality of the laminate is dependent on the skill of the moulder and the laminating conditions.

What is spray moulding?

In this method the moulder uses a spray gun to deposit simultaneously resin and chopped glass rovings. The equipment commercially available varies in detail (Fig. 15), but generally consists of a twin headed spray gun which sprays streams of catalysed and accelerated resin by means of compressed air in conjunction with a glass cutter which cuts the glass rovings and blows them through the resin stream on to the mould. In

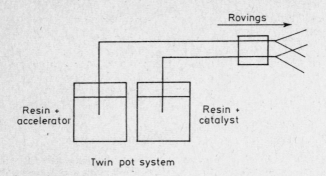

Twin pot system

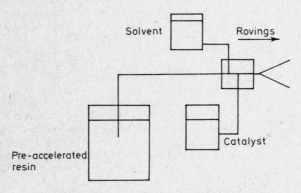

Catalyst injection with internal mixing system

Fig. 15. Spray systems

use the operator traverses part or the whole of the mould with a resin/glass deposit, which is then rolled to complete wetting out and remove air bubbles before the next coat is applied. A skilled operator will complete the whole lay-up in one operation.

30

What are the advantages of spray moulding?

1. The raw material cost of glass rovings is lower.
2. Material wastage is low.
3. The moulding time is considerably less than that required for the hand lay-up method.
4. The resin:glass ratio can be regulated as required.
5. The length and quantity of the glass fibres can readily be regulated.

What are the disadvantages?

1. The control of the distribution of the glass fibres is completely dependent on the skill of the operator.
2. Mechanical breakdowns or blockages result in loss of production.
3. The spray equipment requires regular servicing.
4. Strength of laminate is slightly less than with hand lay-up.
5. High styrene loss, and extraction problems occur.

What types of mould are used?

As GRP possesses no inherent form whilst being fabricated, it is necessary to use a mould to impart the required shape to the laminate. In series production moulding the female type of mould is normally used, in order to give a good external finish to the moulding. Male moulds are often used for the production of tanks etc., where an internal finish is required. Furthermore one-off sandwich construction lends itself to the use of simple batten male moulds; however it has the drawback that the moulding requires filling and sanding to a fair hull surface.

What preparations to the mould are necessary before commencing laminating?

The mould has to be treated with what is termed a parting agent to ensure that the moulding will readily separate from the mould. Hard wax is often used alone. However it is best to

31

use a two-stage release system, whereby the mould surface is first coated with wax and then coated with a PVA (polyvinyl alcohol) solution. To ensure that the mould has been properly coated dyes are added to the PVA to give a coloured film making visual inspection easier. Four to six applications of wax should be applied to a new mould to minimise the chance of stick-up.

What is the procedure for producing a moulding by the contact moulding method?

First the mould surface is thoroughly cleaned and polished, then buffed to a high gloss. The wax release agent is then applied to the mould surface and then polished. At this stage the PVA release agent is applied and allowed to dry. The catalysed gel coat resin is then applied to the mould surface, which may be done either by brush or by spray. When the gel coat is touch-dry (about 40 minutes) the laminating resin is applied, again either by brush or by spray, to a section of the mould surface. The glass reinforcement pre-cut to shape is then placed in position over the resin. Then by means of rollers (Fig. 16) the reinforcement is worked into the resin, making sure that the resin is thoroughly backed up and air

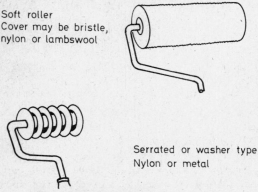

Soft roller
Cover may be bristle,
nylon or lambswool

Serrated or washer type
Nylon or metal

Fig. 16. Typical rollers for hand lay-up

pockets removed. With CSM the resin tends to rise quickly through the mat bringing any trapped air with it. After a short time the binders and sizes tend to dissolve and the reinforcement loses its resilience as the strands bed down and readily knit together at laps and joints. This operation is repeated until the mould surface is covered by the first layer of reinforcement. Subsequent layers are then applied in a similar manner until the required laminate thickness is obtained.

Are there any limits to the number of layers of laminate that may be applied at any one time?

Generally no more than three layers of resin and 600 g mat reinforcement should be applied at any one time without allowing the resin to commence gellation. This avoids the possibility of damage to the gel coat due to the build up of exotherm. Moreover the rapid build up of laminate on vertical and near vertical surfaces can lead to drainage whereby the resin creeps to the lower areas of the mould thereby distorting the resin:glass ratio. Steel or aluminium moulds dissipate heat more quickly and therefore it will be found that more layers than the above may be laid-up at one go.

Does the laying up have to be completed in one operation?

On large hulls it is not possible to complete the process in one operation; however, each successive layer should be completed. The time lapse allowable between the successive layers depends on the curing rate of the resin, and as a general guide, up to 48 hours is acceptable. However if the delay is excessive, the cured layer will have to be abraded to assist the bond with the next layer.

How are the resin:glass ratios controlled in the contact moulding process?

This is achieved by issuing a given weight of resin to the individual laminations for each unit area of reinforcement.

Resin:glass ratios for chopped strand mat are 2¼ or 2½ to 1, and for woven roving 1¾ or 1 to 1.

All glass should be weighed before laying-up to determine the amount of resin required.

When may the internal stiffening be added to the moulding?

This is generally done whilst the hull moulding is still in the mould, the advantages being that it allows the hull laminate a longer curing time in the mould and gives a much stiffer structure which is easier to handle when being removed from the mould.

When may the moulding be removed from the mould?

The moulding must not be removed prematurely as this may give rise to over-stressing of the laminate and possible distortion. The actual time required in the mould depends on the type of resin used. This should be not less than 24 hours or that recommended by the resin manufacturer, after completion of the lay-up. Without the use of a Barcol hardness tester one can only guess that the laminate is cured enough to allow removal from the mould. Recommended degree of cure is 20—25 Barcol hardness.

How are the mouldings released from the mould?

There are many ways whereby mouldings may be released from the mould, such as the use of water which is allowed to settle between the moulding and mould surface, or compressed air, or jacking points, etc. In cases where the moulding sticks, sharp blows to the mould often facilitate removal. However, this should be undertaken with care to avoid damage to both the mould and the moulding.

How long does it take for the moulding to become completely cured?

It is only possible to achieve full cure of the moulding if both the laminating and subsequent maturing period is carried out under controlled conditions of temperature and humidity. When completed the hull moulding should remain in the mould for at least 24 hours prior to being released, and then left for at least 19—21 days in a maintained workshop temperature.

Is it possible to accelerate the curing process after the moulding is completed?

If the moulding is subjected to what is termed 'post heat', faster curing and superior mechanical properties are achieved. In this the moulding is exposed to heat for several hours. The table below is an approximate guide to the relative times required to reach an advanced state of cure.

Temperature ($^\circ$C)	30	40	50	60	70
Time (hours)	30	16	10	6	4

The principal advantage of the system is a quick turnover of the mouldings.

5

WORKSHOP CONDITIONS

Does the moulding of GRP need to be carried out under a controlled workshop environment?

High quality work must be carried out under controlled conditions. Thereby a consistent end product can be achieved which can be guaranteed to have the designed structural properties.

What factors contribute to the controlled environment?

The primary factors which contribute to the workshop environment are: insulation, heating, air conditioning, humidity control, ventilation and dust extraction.

What is the function of insulation?

Insulation assists the heating or air conditioning plants to achieve a satisfactory working temperature, to restrict fluctuations due to external atmospheric changes and/or malfunctions of the heating and air conditioning plants. The extent of insulation required depends on climate, prevailing winds, etc., and degree of shelter afforded by neighbouring buildings.

Why is heating important?

For the correct cure to take place workshop temperature should never be below 13°C (55°F). A temperature of 18°C (65°F) is ideal for proper cure and comfortable for the operator. Higher temperatures will result in premature gelation of the resin. This can be offset be reducing the addition of accelerator but to a limited degree. It is important to maintain proper cure that the shop temperature is thermostatically controlled over the complete 7 days a week and 24 hours a day. No doubt employers are reluctant to do this because of today's heating costs. Consequently the heating plant should be capable of maintaining this temperature range in the coldest weather which is likely to be encountered during moulding operations. A particular point is that in winter the optimum working temperature may not be achieved until later in the morning.

When is air conditioning needed?

In countries where the ambient temperature exceeds $72°F$ $(22°C)$ an air conditioning system in conjunction with a fully insulated workshop is essential. However, in temperate climates the workshop temperature can often exceed the critical level, in which case the warm air can be drawn off by extractor fans or the dust extraction plant.

What is the function of humidity control?

Should the humidity in the moulding area exceed 75%, the glass fibres will pick up moisture which affects both the bond and the cure of the laminate. Consequently in areas where this figure is frequently exceeded, a de-humidifier will need to be fitted in conjunction with the heating and ventilating plant. Where the area is subject to wide fluctuations in humidity the use of air blowers or hot air heaters may suffice. As previously mentioned the workshop may be subject to humidity rises when the plant is shut down, with the consequent harmful effects to the laminate.

Why is ventilation required?

The heavy vapours emitted from the polyester resin are injurious to health, therefore extraction plant must be installed to safeguard the operator. Extraction should take place at a low level. Ideally extraction should take place at source. A certain amount of styrene is lost due to natural evaporation, especially during the actual lay-up procedure, and particular attention should be paid to eliminate cold or warm draughts. Draughts will accelerate the loss of styrene resulting in degrees of permanent undercure of the laminate.

Why is dust extraction necessary?

If trimming is carried out in the moulding area dust extraction plant will be required, as dust can retard and inhibit the resin cure. The systems used can range from a permanent trunked system to small portable vacuum cleaners. Cutting of timber, especially plywoods, and grinding and cutting of GRP components should take place separately from the actual lay-up shop. Certain materials if allowed to mix with the resin will act as inhibitors. To safeguard the health of the operators proper dust extraction must be installed. Again extracting at source is ideal and is usually the cheaper method. Extra precautions must be taken when cutting laminates composed of resins containing trioxides (fire retardants); operators should wear full protective clothing, i.e. masks, gloves, etc.

What arrangements are required for resin storage?

Depending on the viscosity, resins if stored in a cool environment will last from 6 to 12 months. A warm store will reduce this shelf life. The components of a polyester resin are kept from cross linking (polymerising) by addition, by the manufacturer, of an inhibitor. Over a long period of time the inhibitor breaks down allowing the resin to cross link to a certain degree resulting in the resin becoming thick, perhaps even rubbery, and therefore unusable. Resins should be well protected from direct sunlight, and those containing fillers and pigments which are supplied in drums should be turned frequently to prevent them from separating.

What arrangements are required for storage of catalyst MEKP?

MEKP is continuously decomposing and if proper care is taken should not be used after a period of 6 months. Ultra-violet light accelerates the decomposition; therefore catalyst should be stored in a cool and dark place. Catalyst should be stored separately from accelerator. Should the two come directly in contact with one another the reaction would be violent, perhaps explosive. Catalyst or catalysed resin should never

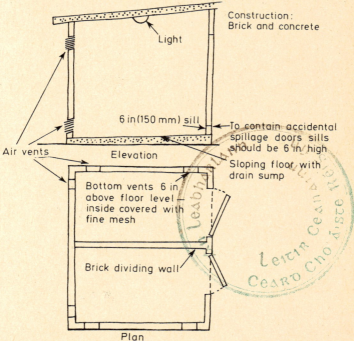

Fig. 17. The resin and catalyst store must meet safety requirements. Lights and switches must be fully enclosed and flashproof and doors one hour fireproof

be put in metal containers. The storage of resins, catalysts, accelerators and acetone cleaners all come under the Highly Flammable Liquids Regulations (Fig. 17).

What arrangements are necessary for resin handling and mixing?

All the materials should be labelled and each allocated its own specific storage area, and preferably given a date to prevent misuse, wrong formulation or use of obsolete materials. The catalysts should be kept apart from the other materials and stored in a flame-resistant store as required by the local government authority and the Highly Flammable Liquids Regulations. The resin, catalyst, and accelerator required for use should be decanted into suitable containers ready for mixing. The quantities of the resin mix will be measured by volume and weight, and the scales used should ideally be direct reading for ease of use, and of robust construction. For smaller quantities of liquids, catalysts and accelerators are more conveniently measured by volume in graduated cylinders (Fig. 18). Catalysts and accelerators should not be measured in the same cylinders and separate equipment should be kept for each. The equipment should be kept as clean as possible, to give accurate quantities and to prevent cross contamination. Excessive mixing of the resin mix should be avoided as this will cause air entrapment in the mix and loss of styrene. When the resin mix is issued to the moulder a wide variety of non-metallic containers can be used: polythene containers are long lasting and easy to clean. However, whatever type is used, a clean pot should be used for every fresh mix.

What arrangements should be used for the storage of reinforcements?

The reinforcements should be stored under clean conditions in a dry atmosphere. The temperature should not be less than 15°C and the relative humidity should not exceed 70%. Most binders are water soluble; water kills the polymerisation, therefore glass reinforcements must be stored in a dry place.

40

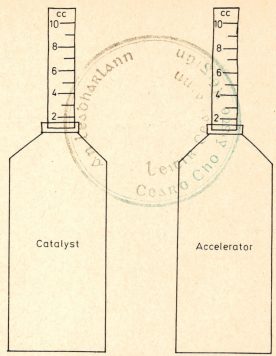

Fig. 18. Unbreakable polythene burettes for safe and accurate measuring and dispensing of dangerous liquids

Shelf life is unlimited. The storage arrangement should be such that all materials are used within their shelf life. The type, weight and date of manufacture should be clearly marked on the wrappings.

What arrangements are necessary for the handling of the reinforcements?

The reinforcements should remain in their wrappings until required for tailoring. Before use they should be exposed to room temperature for 48—72 hours (i.e. the moulding shop) to ensure that they are free from moisture. After cutting the reinforcements should be stored on racks, preferably covered to minimise dust contamination, until required.

How hazardous is the use of GRP materials?

All toxic and inflammable materials present a hazard. The danger is reduced by recognising the hazards, taking no risks, and developing a set of safety rules and procedures which are kept at all times.

What are the main items to consider for safe operation?

1. Hazards of the various ingredients and chemicals.
2. Storage of the various chemicals.
3. Correct mixing procedures.
4. Correct handling of cleaning liquids and solvents.
5. Cleanliness and tidiness of the workshop.
6. Personal hygiene.

In what way is polyester resin hazardous?

It is inflammable. Most polyester resins are classified as highly flammable under the Highly Flammable Liquids and Liquefied Petroleum Gases Regulations 1972. A likely cause of a major fire is the bulk polyester resin being ignited by a fire started elsewhere. Bulk resin stores are therefore best sited in a separate brick building or in underground tanks away from the main workshops. The building should be well-ventilated and have adequate means of fighting a fire, if possible automatic. All electrical fittings should be flash proof.

How should a polyester resin fire be fought?

Preferably with dry powder extinguishers of sufficient capacity, or alternatively with a hosed water supply if there is no danger from contact with the electrical supply.

Are there any other fire dangers in the use of polyester resin?

Yes. Static electricity can be generated in piped supplies, and its discharge could cause a fire. To prevent static build up splash filling should be avoided and pipe lines and containers should be earthed and bonded.

Does polymerisation present a fire hazard?

Yes. In thin sections associated with hull or deck work the risk is slight, but premature polymerisation in a confined bulk of resin may become uncontrollable and could lead to an explosion in the container or the ignition of highly flammable vapours in the area.

Are there any fire hazards in the working of finished products?

The dust generated in machining can form an explosive mixture with air. Dust extraction equipment must be earthed.

What fire risks are associated with catalysts?

Catalysts are organic peroxides and are self-igniting, highly flammable and can be explosive. They are likely to ignite if they come into contact with cellulose materials such as cotton, paper and wood. A catalyst fire should be fought with large amounts of water.

What precaution should be observed in using catalyst and accelerator?

On no account should catalyst and accelerator be allowed to come into direct contact as the reaction is violent and could be explosive. Catalyst and accelerator should be kept in separate stores.

What is the most flammable material used in GRP construction?

Acetone, which is used as a cleaner and solvent. It has a lower flash point than petrol and should be treated with extreme caution. It should be kept in brick stores away from the work area, and must be transferred from store to cleaning bays in non-spillable and flameproof containers.

Because of the high fire risk containers used for cleaning should themselves be stored in a deeper receptacle which will retain the escaping vapour. To conform to the strict regulations relating to flammable liquids the receptacle should be vented outside or connected to the workshop extraction system.

How can the fire risk be minimised in GRP boatbuilding?

By observing all fire regulations. All mixing should be carried out in a mixing bay which must be ½ to 1 hour fireproof and with light fittings and extractors of a flash proof standard. Worktops must be non-metallic, and all containers must be clearly labelled. Sand, water and appropriate extinguishers must be handy.

Only sufficient material for the day's work should be brought from the store, and surplus materials should be returned.

How great is the health hazard in working with GRP?

It is not great provided that the hazards are understood, the rules observed, and all materials are used with care and common sense. First aid should be ready to hand. Particularly in small moulding shops where skilled medical aid might not be immediately available, one operator should understand and be responsible for treatment of minor accidents, and all operators should be encouraged to become proficient at the appropriate first aid. For more serious accidents professional medical aid should be immediately sought.

What are the main rules governing health in the workshop?

1. Protect the skin from contact with all chemicals, and do not wash hands in acetone.
2. Protect the eyes from splashes of catalyst or catalysed resin. Wear goggles when handling large amounts.

44

3. Do not inhale the dust caused by cutting and grinding of laminates, particularly those containing fire-retardant additives, and take special care when handling powder type thixotropic additives. Wear eye shields and masks or dust hoods whenever possible.

4. Do not smoke in the mixing bay or in the vicinity of flammable or toxic materials.

5. Do not eat or drink where dust, powders or toxic chemicals may contaminate food. Eat and drink away from the work area and only after careful washing and removal of overalls.

6. Avoid exposure to the light from all lamps used in the curing process, particularly the high intensity lamps used for surface coating applications.

7. Read and understand all manufacturers' warnings and all health and hazard notices that apply to your own shop.

6

CONSTRUCTION DETAILS

What types of laminates are used in GRP boats?

The structure of a GRP boat may be a single skin laminate, sandwich construction consisting of thin laminates separated by a suitable low density core, or a combination of both types.

What is single skin laminate?

As the name implies, this consists of a hull moulded from a single skin which comprises several layers of laminate, either stiffened or unstiffened.

What is meant by the terms unstiffened and stiffened?

The term unstiffened means that no additional stiffening sections are added to the hull laminate, the strength of the hull being derived from the skin laminate and the hull shape. This arrangement is usually confined to very small craft.

The term stiffened means that the hull laminate is strengthened by the addition of internal stiffeners. This is usually required with larger hulls, or where large flat panels are employed. They prevent the unsupported skin laminate from becoming too flexible. The stiffening is usually arranged in two basic directions, longitudinal or transverse, the choice depending mainly on strength and production considerations.

What is sandwich construction?

In sandwich construction two thin laminates are separated by a light-weight core which increases the rigidity of the panel by increasing the effective thickness without the use of solid laminates (Fig. 19). With the proper application of the sandwich method hulls can be built with a high reduction of weight and no loss of rigidity. Because of the inherent rigidity and lightness

GRP laminate
Core material
GRP laminate

Fig. 19. Normal sandwich construction (top) and cross linked sandwich construction (bottom)

sandwich construction is often used for decks, cabin tops and interior bulkheads. It also readily lends itself to the production of 'one-offs' and prototypes, when foamed plastic is used for the core. A large number of craft, including large side wall hovercraft, have been constructed using this method. Instead of the normal female mould, a male mould is used which can be simply constructed from battens. Sheets of the core material are then laid over the moulds and temporarily fastened in position. The external laminate skin is then applied to the foam. When the laminate is sufficiently cured it is removed from the mould and turned right side up. The internal laminate is then applied.

What materials are used for the core?

Many cores are available, i.e. GRP honeycomb, aluminium honeycomb, PVC, polyurethane, and other plastic foams and end grain balsa. Each of these cores has its own characteristics and properties enabling the builder to choose the one most suitable to his needs and requirements of the end product.

47

What is composite construction?

This generally refers to the fitting of a wooden deck and superstructure to a GRP hull, or the encapsulation of wood or metal structural members into a GRP hull. This mode of construction although sometimes used for prototypes can cause difficulties due mainly to the dissimilarities of the properties of the materials used.

What arrangements are used for the fabrication of keels?

The keel is formed by laminating additional reinforcement along the centre line of the hull, and extends from the stem to the transom. It can be of constant width throughout or reduced in either width or weight or both towards the ends.

Left and right hand reinforcement is carried over alternately to create laminate increase.

Fig. 20. Keel reinforcement formed by overlapping bottom layers

Increase in laminate weight is tapered off to bottom weight at approx 25 mm per 600 grammes

Fig. 21. Keel formed by additional material

It may be formed by overlapping the bottom laminates (Fig. 20) or by additional layers (Fig. 21). In sailing vessels with deep fin keels access to this area in the mould is awkward, thus it is common for the hull to be moulded in two halves. The mould is then bolted together and the laminating is completed by adding additional material at the joint (Fig. 22).

48

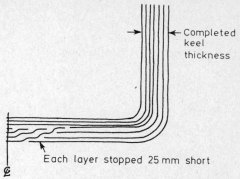

Each layer stopped 25 mm short

Fig. 22. Keel formed by moulding hull in separate halves

How are ballast keels attached to the hulls of GRP sailing yachts?

Ballast keels can be fitted either externally or internally (Fig. 23). With an external ballast keel the major problem is fit. Ideally the keel should fit close enough for just a thin layer of sealant. However, the castings are normally slightly

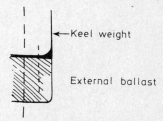

Fig. 23. External ballast keel

undersized and usually require packing to obtain a close fit. If thick layers of flexible packing are used, the keel can rock about the fastenings with resultant leaks. Thus the flexible sealant should be kept to a minimum and the remainder built up with resin based filler. The sealant should be of a type

49

which remains permanently flexible; those suitable are poly-sulphide or silicone based. An alternative is to use a neoprene gasket but this requires a uniform gap to be effective.

The external ballast keel is attached to the hull by means of keel bolts which are usually fabricated from stainless steel or monel. As these bolts are heavily loaded they need to be fitted with large washers or plates to distribute the load over as large an area as possible.

How is internal ballasting achieved?

Internal ballast keels are generally as shown in Fig. 24. The ballast may be in either one piece or more often loose pieces of scrap metal, punchings or shot, set in resin or concrete.

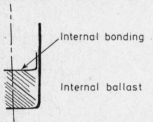

Fig. 24. Internal ballast keel

When resin is used as a binder it should be built up slowly to avoid any damaging effects of exotherm. Loose ballast is not recommended as it can work in a seaway and damage the internal surface of the hull laminate. In the case of deep keel sailing boats the GRP hull has enabled the ballast keel to be positioned inside the hull. This has two main advantages:

1. No water leakage or breakdown likely at the hull keel join.
2. No keel bolts to be corroded through electrolysis caused by dissimilar metals.

The hull is increased in thickness to accommodate the ballast keel and any damage incurred through probable ground-ing is easily repaired and presents no problem. The advantage of the external ballast keel is that it and not the GRP shell absorbs the knocks and abrasions and as it is virtually indestructible little harm is done. However this is offset to an extent by the initial cost, the need for keel bolts, and the extra maintenance required.

With internal ballast, the initial cost is low, though knocks and abrasion are taken directly by the GRP hull, which tends to wear quickly. Furthermore the internal surface and centre-line joint remains inaccessible. However, a number of yachts have been designed with protective shoes of wood or GRP specifically intended to absorb the wear, leaving the main fabric of the hull undamaged.

What arrangements are used for hull stiffening?

As stated earlier, in very small craft the stiffening is derived from the compound curves of the hull shape. With vessels in the intermediate range additional hull stiffening may be derived largely from fixed internal joinery, such as bulkheads, bunksides, etc. However, in larger sailing yachts and power craft, the internal stiffening arrangements are independent of the internal joinery. This, while being expensive, gives complete flexibility to the internal arrangement.

In the case of a fast power boat the stiffening would be longitudinal to give good flexural strength to the hull. A heavy work boat on the other hand would need stiffening in the form of intercostals and frames to give good transverse strength also. On large motor boats it is conventional for the fabricated engine beds to extend from the transom to right forward to give added stiffness.

What sections are used for stiffeners?

Typical arrangements are shown in Fig. 25, which shows a simple arrangement which is built up from layers of unidirectional

Unidirectional fibres

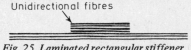

Fig. 25. Laminated rectangular stiffener

51

fibres. This arrangement is often used in areas where both the span and depth of the stiffeners is restricted. Where larger spans and loads are involved, this type becomes uneconomical in material and is replaced by those shown in Fig. 26, which consists of hollow or solid cores around which layers of laminate are formed. Solid cores are normally of foamed plastics and

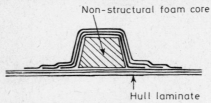

Non-structural foam core

Hull laminate

Fig. 26. Foam core stiffener of 'top hat'
section

are intended to give the desired profile to the laminate and are not considered as contributing to the strength of the stiffener. This also applies when timber is used for the core. Hollow cores may be fabricated from expanded sheet metal, cardboard or plastics. The stiffness and strength of the sections may be varied by adjusting the section depth and the laminate weight and also by incorporating unidirectional fibres or carbon fibres (Fig. 27).

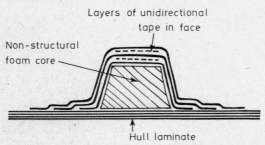

Layers of unidirectional
tape in face

Non-structural
foam core

Hull laminate

Fig. 27. Foam core stiffener with unidirectional tapes

52

How are the stiffeners ended at the deck edge or at bulkheads?

Fig. 28 shows how the stiffeners can be tapered off at their ends. To ensure continuity of strength the laminate should be locally increased in thickness to compensate for the loss of strength. Furthermore the stiffener laminate should be neatly run into the hull laminate to avoid any local discontinuity.

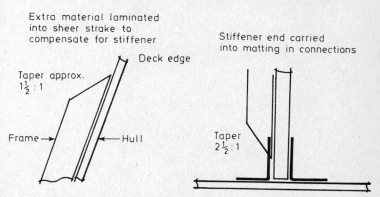

Fig. 28. Stiffeners are tapered off at their ends as in this side frame at deck edge (left). The drawing (right) shows the attachment of a bulkhead stiffener

What materials are used for bulkheads and how are they attached to the hull?

The most common material used is marine plywood. However, bulkheads can be fabricated in sandwich construction although this is considerably more expensive. The bulkheads are normally unstiffened in small craft and it is only in the larger sizes that stiffening is required to prevent buckling or vibrating. The corners of any openings need to be cut on a radius to avoid stress concentrations and possible fractures when heavily loaded. The matting-in connections are generally as shown in Fig. 29, the thickness being increased for watertight bulkheads and those

53

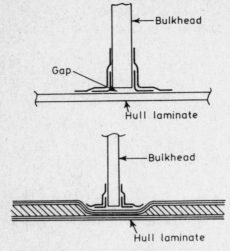

Fig. 29. Typical connection between bulkhead and shell for single skin hull (top) and sandwich construction (bottom)

in way of masts. It is preferable to fit the bulkheads while the hull is still in the mould as they can be fitted accurately and a better bond can be obtained. It is good practice to leave a small gap between the bulkhead and the hull, or to place a foam cored stiffener between the two, as this will eliminate hull distortion due to the shrinkage of the bonding laminate as it cures.

What arrangements are used for engine seatings?

Installations suitable for use with low power as found in sailing yachts and small motor cruisers are shown in Fig. 30. As an alternative pre-formed GRP sections can be used for the engine girders. With larger engines it is usual practice to fit steel bearers through-bolted to the girders. Ideally these should be

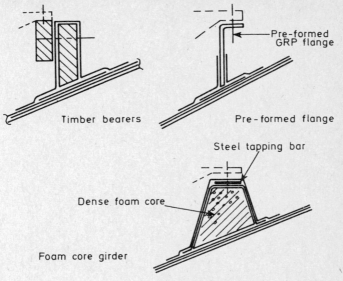

Timber bearers

Pre-formed GRP flange

Pre-formed flange

Steel tapping bar

Dense foam core

Foam core girder

Fig. 30. Engine girders

as long as possible. The girders should extend the length of the machinery space and be fitted with floors and brackets. Fig. 30 shows an arrangement often found in the largest installations, consisting of GRP laminate formed over a heavy density core, which should be completely integrated into the hull stiffening arrangement.

May fuel and water tanks be fabricated from GRP?

Although rarely used in the early years of GRP boat construction, such items are now commonplace. GRP tanks have a number of advantages: they do not rust or corrode, are non-magnetic and are simply constructed. However, the disadvantage is vulnerability to damage. Both types of tank may be integral with or separate from the hull structure. With fuel tanks, a separate installation is probably safer than the integral type in view of the risk of fuel leakage due to hull damage. However the risk can be minimised by careful positioning of the tanks away from more vulnerable areas. The tanks are normally fabricated on separate moulds, such that the gel coat will form the internal surface of the tank. When the laminate is complete, stiffening and internal baffles may be added. The completed moulding is then bonded into the hull. The internal bonding is then completed via access through manholes (Fig. 31).

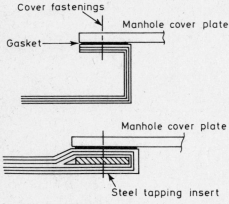

Fig. 31. Manhole details

The internal surface of the tank should be given a heavy resin coating, and if the tank is being used for drinking water, it is important that the resin should be of a type that does not contain additives other than the catalyst and accelerator, as certain additives have been known to taint the water.

Special resins and gel coats are formulated by the resin manufacturers specifically for GRP food and water containers. Information regarding the decontamination procedure of newly constructed containers is also available from the resin manufacturers.

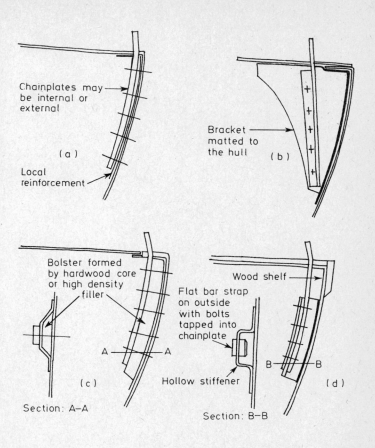

Chainplates may be internal or external

(a)

Local reinforcement

Bracket matted to the hull

(b)

Bolster formed by hardwood core or high density filler

Flat bar strap on outside with bolts tapped into chainplate

A —— A

(c)

Hollow stiffener

Section: A–A

Wood shelf

B —— B

(d)

Section: B–B

Fig. 32. Details of chainplate attachments to hull showing chainplate (a) through deck fastened to bulkhead or bracket (b) bolted internally to hull (c) through deck attached to bolster (d) attached to 'hat' section to avoid bolting through hull

57

How are the chainplates attached to the hull?

Fig. 32 shows a number of methods of attaching chainplates. The plates should be of ample size and well fastened to the structure to give adequate load distribution. Where the chainplate or chainplate fastenings penetrate the hull and deck laminate, they should be watertight with a flexible sealant rather than with resin which cracks under the rigging loads, causing leaks.

What fabrication techniques and arrangements are used for the construction of decks and superstructures?

The most common arrangement is that of a GRP deck and superstructure moulded as one unit. However this precludes individual deck arrangements which are best fabricated from timber, as the cost of producing moulds to individual requirements is expensive. A compromise is often made by using a GRP deck with a timber superstructure. The structure of a GRP deck moulding may be either a single skin construction throughout or a combination of sandwich construction on the horizontal surfaces and single skin on the vertical (Fig. 33).

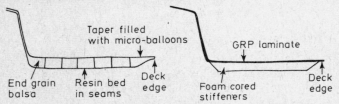

Fig. 33. Balsa wood sandwich deck (left) and stiffened single skin laminate deck (right)

One of the more popular core materials is end grain balsa wood. To construct a deck using this material, the outer face laminate is laid-up in a female mould in the normal fashion. Small blocks of the balsa wood are bedded on to the laminate with resin. The exposed end grain is sanded flush and sealed

58

with resin. The inner face laminate is then laid-up over the balsa wood. An alternative method using a foam core is to lay alternative strips of the foam which are then matted into the outer laminate. The remaining spaces are filled with the foam, which is then covered by the inner laminate. Additional reinforcement should be laminated in the structure at deck openings and at areas where there is a pronounced change of shape.

What arrangements can be used for the deck to hull connection?

Fig. 34 shows a number of typical deck to hull connections which have proved successful. Ideally the connection should fulfil the following criteria:
1. The joint should be waterproof.
2. The joint should be such that it does not require great accuracy of fit to avoid stressing the laminate.
3. The strength of the joint should be such that it is equal to the weaker of the two pieces being joined.
4. The joint should be of a simple design to facilitate easy moulding and inspection.

May metal fasteners be used with GRP laminates?

A number of different types of fasteners have been used effectively in GRP laminates; these include bolts, rivets and screws. Due to the marine environment, they should be fabricated from corrosion resistant materials such as bronze or stainless steel, or treated with a corrosion resistant coating such as galvanising. Ideally the laminate should be increased in weight where metal fasteners are used. As an approximate guide, the fastenings should not be spaced closer than 3 diameters and no less than 3 diameters from the face edge of the laminate. Also the holes for the fasteners should be coated with resin to prevent the ingress of water into the laminate.

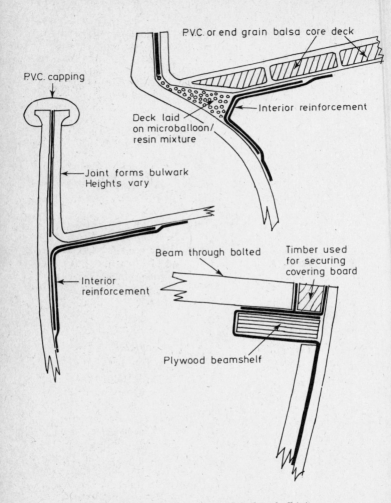

P.V.C. or end grain balsa core deck

P.V.C. capping

Deck laid on microballoon/resin mixture

Interior reinforcement

Joint forms bulwark Heights vary

Beam through bolted

Timber used for securing covering board

Interior reinforcement

Plywood beamshelf

Fig. 34. Sections showing typical deck to hull joints

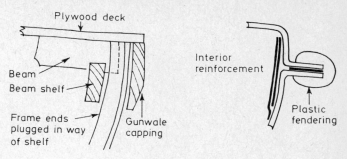

Plywood deck

Beam

Beam shelf

Frame ends
plugged in way
of shelf

Gunwale
capping

Interior
reinforcement

Plastic
fendering

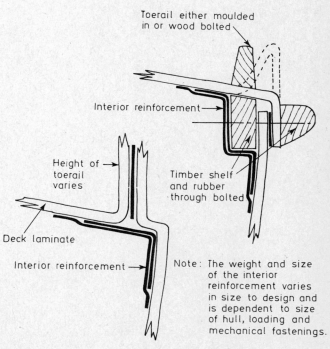

Toerail either moulded
in or wood bolted

Interior reinforcement

Height of
toerail
varies

Timber shelf
and rubber
through bolted

Deck laminate

Interior reinforcement

Note: The weight and size
of the interior
reinforcement varies
in size to design and
is dependent to size
of hull, loading and
mechanical fastenings.

What precautions are necessary when using bolt fasteners?

Bolts should be fitted with washers approximately 2½ times the shank diameters. This is to avoid crushing the laminate due to over-tightening. Also the shank diameter should be at least equal to the thickness of the surrounding laminate. The bolt needs to be a fairly tight fit in the laminate, as this will reduce the tendency for the bolt to work in its hole with the attendant possibility of failure.

What precautions are necessary when using rivets?

Again, washers or plates should be fitted under the head and points of rivets to prevent the laminate from being crushed. The rivets should not be placed nearer than three times the diameter to the free edge of the laminate.

What precautions are necessary when using screw fastenings?

Due to the incompressible nature of GRP wood screws should not be used. However, self-tapping screws may be used where the components to be joined are relatively lightly loaded. When used in conjunction with light laminates a metal tapping strip will be needed.

How may fittings be attached to the hull laminate?

With single skin laminates, fittings may be through bolted, or may be bonded and matted in. With through bolting, the holes should be sufficient to give a tight fit to the bolts, and the cut edges of the laminate should be sealed with resin. Fittings such as bollards and cleats, which take heavy loads, should be bedded down on a gasket or a flexible sealing compound. Ideally the laminate in way of such fittings should be increased in thickness. Additionally a backing plate should be fitted. When required, drilled and tapped plates can be moulded into the laminate, to take fittings. The plates should have as large an area as possible in contact with the laminates, to ensure that the loading is evenly distributed. Keying of the plates to the laminate can be improved by heavily scoring or indenting the surface of the metal. A degree of caution is needed in the use of such plates as some steels are subject to corrosion when encapsulated, which will break down the laminate to metal bond causing failure. Furthermore, differences in the physical properties may have the same effect. With foam sandwich construction, through bolting fittings will cause the normal core material to compress. To overcome this problem, the core may be cut away and the laminate reduced to a single skin. Alternatively a material such as plywood may be substituted for the core, which will readily accommodate the compressive loads.

7

SURVEY AND INSPECTION

What factors determine the quality of construction?

One of the most important factors concerning construction in GRP is the workshop conditions and temperature control. Lloyds have found this to be one of the major factors in achieving proper cure in a laminate. It is important that weighing and mixing of ingredients are strictly adhered to as per recommendations.

What survey procedures can be carried out during the manufacturing stage in GRP boat construction?

The manufacturing process may be divided into three parts:
1. Pre-moulding operations.
2. Moulding process.
3. Post-moulding operations.

Each of these stages should be subject to inspection. As the survey procedures for GRP boatbuilding rely largely on visual or non-destructive inspection, a close check needs to be kept on the entire fabrication process to ensure that it complies with the prescribed conditions.

What aspects will be subject to inspection during the pre-moulding operations?

The area covered by the pre-moulding operations comprises the following: material, the laminating sequence and the mould.

Consider the materials: the components of the resin system,

the accelerator/catalyst content, and the pigment content, should comply with the design requirements. Ensure that the proposed proportions will give the correct results. With regard to the reinforcement, ensure that the type and weights comply with the design requirements and that the glass type is the correct one for the purpose.

When considering the laminating sequence, the proposed procedure and phasing should be examined. Changes in the laminate thickness, overlaps and the sequence of the various types of laminate need also to be considered. Areas where pronounced change of shape occurs in the mould can cause laminating problems and should be discussed before moulding. If new laminate arrangements are being employed, this should be discussed with the moulder.

Finally the mould should be checked to ensure that it is properly supported, and if the mould is split along the centre-line, that it is correctly aligned. The mould surface should be examined for local unfairness, due to incorrect support, and for blemishes and minor damage. Furthermore the position of the mould in the workshop with regard to lighting, heating, ventilation and dust should be satisfactory.

What aspects will be subject to inspection during the moulding operation?

A close check should be made to ensure that the laminate is being throughly inpregnated and consolidated; also that the phasing of the operation is being maintained. A sample of the resin should be examined for impurities, and the reinforcement should be similarly examined for dirt, etc. To confirm that the resin:glass ratio is being maintained the relative quantity of materials used should be checked.

What aspects will be subject to inspection during the post-moulding operation?

Before releasing the moulding from the mould ensure that the laminate has reached a sufficiently advanced state of cure;

also, that any temporary bracing if needed, is ready to be fitted. Furthermore check that any equipment required for the release will not inadvertently damage the mould or the moulding. After release from the mould the laminate should be examined for surface flaws and such areas marked for further attention. It should be ensured that the moulding is not subject to distortion due to poor supports and bracing.

What problems can be encountered with the resins and reinforcements used for fabrication?

A number of problems can arise with the resin systems and are mainly concerned with deviation from the manufacturers' recommendations. For example, resins have a shelf life of about six months if stored under the correct conditions. Any resin used whose age is in excess of this period should be treated with caution and should be subjected to a laminating test.

Many errors can arise from poor workshop hygiene and this can lead to incorrect mix, cross contamination and foreign matter such as dirt and dust in the resins. Obviously these can be largely eliminated by strict control over the handling, mixing and dispensing operations. With reinforcements, one problem that sometimes occurs is that of uneven or poor quality areas; with inspection these areas can be readily detected and removed.

Like resins, reinforcements tend to attract dirt and dust in the workshop, and again these problems can be eliminated with good storage and handling arrangements.

How may the laminate thickness be assessed?

Calipers may be used on areas of the moulding which are readily accessible. Those areas which cannot be reached can be assessed by examining cut outs from hull fittings or by drilling small holes in those areas required and then examining the thickness, the holes then being plugged. Alternatively ultrasonic meters may be employed, but these are quite expensive.

A comparison between the measured thickness and the known numbers of types of plies of reinforcements will readily give an approximate guide to the resin:glass ratio. However, to determine accurately the resin:glass ratio, a sample of the laminate is taken, and is then burnt; by weighing the sample before and after burning, the resin:glass ratio can be obtained.

How may the laminate quality be assessed?

The tests that may be employed are divided into 'non-destructive' and 'destructive'.

Non-destructive tests are simple and include:
1. Inspecting the laminate with a powerful light source to gauge its consistency and the presence of any internal defects such as voids.
2. Sounding the laminate with a hard metallic object. A clear sound indicates a good laminate. A muffled response indicates that either under-cure or poor impregnation is present in the laminate.
3. Feeling the laminate. If any tackiness is present it will indicate under-cure of the surface laminate.
4. If the characteristic smell of styrene is present, it could indicate insufficient cure or under-cure of the laminate, depending on the date of lay-up.
5. None of the above methods is completely satisfactory in determining degree of cure. The only way to determine the degree of cure is with the aid of a Barcol hardness tester.

The destructive tests available have to be carried out within the confines of a laboratory. These tests are best implemented on samples taken from hull cut outs, and include various standard mechanical and chemical tests from which the characteristics of the laminate can be evaluated. Finally, it should be ascertained that any post-cure requirements have been complied with.

67

How may the degree of cure be assessed?

Once initiated the polymerisation reaction may continue for a considerable length of time, and should this process be interrupted the resultant laminate will be under-cured with attendant reductions in the mechanical and physical properties. Unfortunately there is no satisfactory method of determining the state of cure. The only available means at present is the Barcol hardness tester which measures the surface hardness of the laminate. The instrument is portable and easy to use and is convenient to use on small items.

What are the laminate defects found in GRP?

The more common defects are listed below:

Voids   Voids or air bubbles can occur on the surface or within the laminate. They result from poor wetting out, air entrapment in the reinforcement, or air inclusion in the resin during mixing. Small concentrations of voids or air bubbles in the laminate cannot be avoided, but with careful laminating they can be kept to an acceptable level. However, should they be present in large quantities, the physical properties of the laminate will be impaired.

Blisters   These are due to local delamination of the laminate and are generally caused by entrapped air or solvent. If they are present on the surface over a large area, it indicates that the rate of cure is too fast. If they appear within the laminate it can be caused by insufficient impregnation of the reinforcement with the resin.

Leaching   This is the loss of resin from the laminate exposing the reinforcement. It is caused by the prolonged weathering of under-cured laminate.

Delamination   Delamination is the failure of the inter-laminar bond. This may be attributed to excessive exotherm, contaminated reinforcement or shrinkage during cure. It may also

68

be due to excessive loading, such as impacts or distorting the laminate. Its presence is indicated by opacity in the laminate.

Crazing   This is an indication of localised bond failure between the reinforcement and the resin. If present on the surface it appears as fine cracks, which are normally found in resin rich areas and can have a number of causes. These include incorrect resin mix, excess styrene, insufficient plasticiser or localised over-stressing. When found within the laminate they are generally attributed to excessive humidity during fabrication, contaminated reinforcements, or resin failure due to excessive exotherm.

Dry patches   These are found on the internal surface of the laminate and are indicated by a whitish fibre pattern. They are caused by poor resin distribution, insufficient resin and incorrect surface treatment of the reinforcement. They may be regarded as an indication of poor fabrication technique.

What are the surface defects found in GRP?

The more common defects encountered in the gel coat are listed below:

Poor adhesion of the gel coat   This can be caused by unsuitable choice of resin for the gel coat which cures too quickly, contamination of the gel coat before the resin is applied, or poor consolidation of the laminate on to the gel coat.

Poor wetting of the gel coat   This has the appearance of cavities and depressions on the surface and is generally known as 'fish eyes'. This is normally caused by excessive release agent containing silicones or a gel coat resin with a low viscosity.

Wrinkling   This may be caused by under-cure of the gel coat due to the excessive loss of styrene by evaporation, under-cured gel coat which was subject to attack by the styrene in the laminating resin, or under-cured gel coat which is too thin.

Pinholes   These can be attributed to any or a combination of the following: dust in the gel coat or release agent, air in the resin, air bubbles on the surface of the gel coat, or the gel coat not wetting out on the release agent.

Fibre pattern   This is characterised by the underlying laminate weave pattern showing through the gel coat. It can be caused by insufficient gel coat, the reinforcement being laid-up before the gel coat has cured, or removal of the laminate from the mould while under-cured.

8

MAINTENANCE AND REPAIR

Is GRP a maintenance-free material?

GRP like any other material, needs the correct protection throughout its service life. It is certainly a low maintenance material but usage will ultimately affect the degree of maintenance required.

What can be done to protect GRP against weathering?

Weathering is a slow process, which eventually if unchecked erodes the surface of gel coat making it rough and pitted and subject to accelerated attack. Eventually the gel coat will become porous, thus no longer protecting the underlying laminate. This process normally takes many years. However, if the gel coat is of poor quality or subject to excessive atmospheric pollution it can occur relatively quickly. The most effective safeguard is regular waxing of the gel coat, which then acts as a barrier between it and the weather. Furthermore the wax also provides a measure of protection against scuffs and abrasions and sea-borne oil.

What type of wax may be used with GRP?

The choice of polish is important. Obviously it must be weather resistant, but considering the eventuality of repair work and ultimately repainting, the polish needs to be of a type which can be easily removed. The choice lies between two types, silicones and wax. Silicone polish readily gives a high gloss finish which is water repellent. However, it has

71

a major disadvantage in being extremely difficult to remove. Wax polish is thicker and easier to remove, but requires more buffing to produce a high gloss.

What are the main causes of gel coat deterioration?

Apart from general weathering, the most common cause of minor gel coat damage is abrasion from the rubbing of fenders or mooring warps. Boats lying to trots or tied up alongside other craft at crowded pontoons are particularly prone to abrasion from the continual rubbing motion between vessels. Even new fenders will eventually lose their smooth surface and abrade the gel coat, while rough or badly worn fenders (particularly car tyres and old-fashioned rope fenders) will remove the smooth surface of the gel coat very rapidly.

Bad anchoring practice is also a cause of trouble, and the well-handled craft should always be allowed to lose way through the water before the anchor cable is paid out. Wind against tide conditions will tend to drive a tide-rode (usually deep keeled) boat over her chain, causing rapid damage to topside gel coat. Plastic pipe can be used to protect the topside from the chain, though in less severe wind conditions a bucket streamed astern will tend to prevent the craft from riding over its chain.

How is loss of gel coat gloss rectified?

If abrasion is only superficial it can be polished out using one of the proprietary boat maintenance products, though mild cutting compounds and polishes sold for car refinishing have been found satisfactory.

If the abraded gel coat includes some areas of deeper abrasion which have not actually penetrated the gel coat, a very fine wet and dry paper can be used initially, to be followed by the above procedure.

Should a GRP hull be painted?

If properly cared for, and excepting the possibility of a disfiguring repair, a GRP hull should maintain its appearance for five years or more, depending on the extent and conditions of use. A well maintained sailing dinghy, for example, may never need painting, assuming the owner has the sense not to drag his craft along the foreshore or across a sandy beach. A workboat or fishing boat, on the other hand, is bound to suffer wear and tear and should be repainted as much for protection as for appearance.

What is the procedure for painting the topside of a GRP boat?

Follow the paint manufacturer's instructions implicitly. Paints for GRP hulls have been carefully formulated and generally need more careful application than do those for wooden or steel hulls. Also the more perfect surface of a GRP hull accentuates any faults in paint application, emphasising brush marks, runs and 'holidays'.

The first essential is a perfectly clean surface, with all traces of wax polish removed. Particular care should be taken to remove stubborn silicone waxes if these have been used, as well as traces of varnish dropped from toe rails or rubbing strakes. Do not use strong abrasives, and on no account remove varnish or conventional paint with paint stripper, unless it is one of the specially formulated kind that will not attack the gel coat.

The hull is best cleaned with water and mild detergent, and more persistent marks removed with meths or white spirit.

The topside can then be mechanically abraded using a fine grade of wet and dry paper used wet (with a little detergent to stop clogging), or a key for the paint can be provided by using an etching primer.

What types of paint are available?

The main types available are:
    Conventional yacht enamels usually based on an alkyd resin
    Single-pot polyurethanes
    Two-pot (catalytically cured) polyurethanes

What are the advantages of the particular systems?

Conventional yacht enamels are cheaper and easy to apply. Because they are slower drying brush marks can be worked out more easily and the risk of a dry edge is minimised. They are, however, softer, less resistant to abrasion and generally less durable. A conventional enamel will usually require re-painting after two or three years. One-pot polyurethanes offer a good compromise, are harder and tougher than enamels, and relatively quick drying. Most dry by reaction with moisture in the air and will dry well on calm, humid days (when dust is not raised by ground and wind). A good painting technique must be mastered, and the painter should work in vertical strokes from one end of the hull to the other, without stopping. Avoid overloading the brush as runs can rarely be brushed out, and ensure that each successive stroke meets the wet edge of the previous. A dry patch, or 'holiday', cannot easily be rectified.

Two-pot polyurethanes give the greatest strength, resistance to salt water and chemical attack, and impermeability. They will also give the finest finish if applied with skill. Two-pot polyurethanes can be flatted off and buffed to produce a very fine finish, free from all traces of brush marks. Like one-pot polyurethanes, these paints are quick drying, and a similar technique should be adopted. Drying rate, however, depends not on humidity but temperature.

Why must GRP boats be painted below the waterline?

Firstly, to prevent marine organisms such as barnacles and weed attaching themselves to the hull and thereby increasing the resistance of the hull to movement through the water, and secondly, to form a relatively impervious layer over the gel coat, and so reduce the possibility of water absorption by the hull and consequent osmotic blistering. Special paints with low permeability and antifouling qualities are developed for under-water use.

What is the procedure for painting a newly moulded hull below the waterline?

Before painting the hull sufficient time must be allowed for the hull to cure (about 3 weeks at 60°F) and any surface imperfections, blisters, etc., must be removed and filled.

The hull is then mechanically abraded using abrasive wet and dry paper, or other process which will key the surface of the gel coat without removing an excessive thickness.

Many different materials have been evolved for painting below the waterline, and it is essential that the paint manufacturer's instructions are carefully followed. One method is to prime the surface with two-pot polyurethane thinned by the addition of 15% thinners. Subsequent coats are then applied unthinned until an adequate thickness has been built up.

Alternatively, undercoats and antifoulings based on epoxide resins can be used. These have a very high resistance to porosity, and generally require fewer coats than polyurethane.

If the traditional type of porous 'soft' antifouling is to be used, sufficient coats of an approved underwater undercoat must be applied to protect the gel coat from ingress of water.

How should the bottom be re-painted?

Loose antifouling should be removed if the craft is already painted. This is best done with a coarser grade of wet and dry paper, used with a plentiful supply of water. Rubber gloves should be worn, or the hands protected with barrier cream. Goggles, mask and hat should also be worn and the face and arms thoroughly washed as soon as the job is completed, and certainly before food is taken. Antifouling compounds contain poisons which present a real hazard and are a frequent cause of dermatitis, eye damage and respiratory troubles.

When the antifouling is cut back to sound paint, a further coat or two can be applied — but only if the new material is compatible with the existing antifouling. If in any doubt consult the manufacturer or cut back the old antifouling to undercoat or primer.

How may minor surface damage to the moulding be repaired?

Minor surface damage includes damage to the gel coat, such as scratches, cracks and cavities.

Before applying gel coats or glass/resin layers to a repair the area to receive the repair layers must be abraded, not only to get rid of any contaminations but also to provide a key. The one operation of abrading accomplishes this.

If the damaged area is small, application of gel coat alone may suffice. Otherwise resin putty will be needed to build up the surface, which can then be finished with gel coat. Whilst it is possible to make effective repairs with pigmented resins, there are limitations. All colour pigmented resins fade, making it difficult to provide a good colour match.

How may ruptures to single skin laminates be repaired?

This is the easiest major repair, and the procedure is as follows. The damaged edges of the laminates are cut back until the

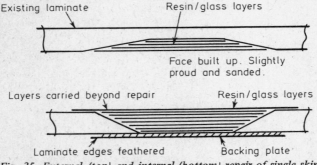

Fig. 35. External (top) and internal (bottom) repair of single skin laminate

sound material is reached, ensuring that any corners are well rounded. The cut edges are then feathered back as in Fig. 35. The exposed laminate is then thoroughly cleaned and abraded. A backing plate of Formica, sheet metal or similar is coated with a release agent and then temporarily secured over the hole. The repair is completed by laying over the hole successive layers of reinforcement and resin, until the repair equals the thickness of the surrounding laminate. At this stage further layers of reinforcement and resin are applied that extend in all directions beyond the repair. Finally a piece of polythene film or similar is placed over the repair to protect against under-cure. Repairs on single skin laminate are normally carried out from inside. Satisfactory repairs can be achieved by laying-up from the outside in which case the procedure is the reverse of that previously described. However, this method does have the disadvantage that the surface requires fitting and sanding to produce a fair finish.

How is a more complicated repair made?

If the repair is in an area of complex shape the fault must first be temporarily repaired. Splits, cracks and holes should be filled with a stopper or filler that is easily cut back. The area is then faired in to the original, waxed, and a release agent applied.

When the surface is properly repaired a GRP mould is made of the area. When properly cured the mould is removed and its moulded surface waxed and prepared as above. It is at this stage that the area needing repair is cut out, the edges bevelled and then the mould replaced and the repair carried out as described above.

How may ruptures in thick laminates be repaired?

With thick laminates it is normal to use the double scarf (Fig. 36) whereby the repair is built up from both sides.

How may a blind panel be repaired?

In some boats it is impossible to reach the inner face, in which case the procedure is as follows.

77

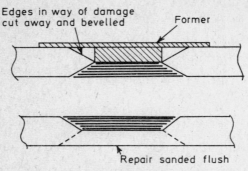

Fig. 36. *Repair of thick laminate with resin/glass layers built up against former (top) and former removed and resin/glass layers built up in recess (bottom)*

The damaged areas of the laminate are cut back as before. A backing plate is then prepared with the addition of two small holes at the centre. Three or four layers of reinforcement are then laid-up on the plate. At this point a length of strong wire is passed through the two holes in the plate so that the ends project through the outer face of the laminate. The facia plate and laminate are then passed through the cut-out and are pressed back into position so that the wet laminate is pressed hard against the interior of the hull (Fig. 37). Tension is

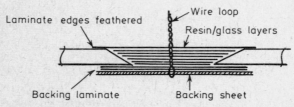

Fig. 37. *Repair method for blind panels*

applied to the wire to maintain this contact as shown in the diagram. The repair is finished as previously described. When the laminate is cured the wires are trimmed back as close as possible. Further layers of laminate are then added. When finally cured the repair is filled and sanded to achieve the required finish.

Alternatively, a backplate can be temporarily attached using self-tapping screws, and removed when the repair is sufficiently cured.

How are ruptures in sandwich construction repaired?

The basic repair is as shown in Fig. 38. The damaged core and laminate is cut away. The outer face is built up against a backing plate; the core is then replaced with bond material which is used as a form for the remaining skin laminate.

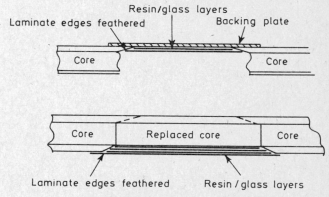

Fig. 38. Repair of sandwich panels

What is osmosis, and what is osmotic blistering?

Osmosis is the process where moisture is absorbed through microscopic pores into what might appear to be impermeable materials. Impermeable as a gel coat might appear it will permit water to pass through it, and over an extended period of time (say 2 or 3 years) this water becomes trapped within the more permeable resin/glass structure of the hull. As with moisture trapped beneath a layer of paint, blistering of the gel coat then occurs. This is not only unsightly, but impairs the structural integrity of the hull.

How is osmotic blistering rectified?

Blisters must be removed and the hull allowed to dry thoroughly. Individual blisters can be ground away to sound material, thus removing all traces of blisters and unsound gel coat.

The hull must be allowed to dry out thoroughly if recurrence is to be avoided, and this may take between three and six months in warm, dry conditions.

Affected areas must be thoroughly filled, or if the complete gel coat has been sand blasted, the surface must be prepared and painted with 3 or 4 coats of epoxy or 6 to 7 coats of two-pot polyurethane. Both epoxy and polyurethane are less permeable than the gel coat and this procedure, followed by antifouling, should prevent recurrence of osmosis.

At the time of writing the causes of osmosis are still being researched, and new materials are being developed that could much reduce the likelihood of osmotic blistering occurring.

GLOSSARY

Accelerator	A substance which increases the hardening rate of a synthetic resin.
Catalyst	A chemical compound which alters the speed of a chemical reaction such as polymerisation without undergoing any permanent change. Curing of polyester resins used in boat construction is not practicable without a catalyst.
Chopped strand mat	A lightly compressed mat formed from continuous filaments of glass which have been 'chopped' to discrete lengths.
Curing	The process of hardening a polyester resin.
Delamination	The breakdown of the structure of the laminate by the separation of the layers.
Exotherm	The build up of temperature which results from a chemical reaction, in the case of GRP between catalysed resin and accelerator.
Female mould	An indented, or concave, mould.
Gel coat	The thin layer of unreinforced resin on the outside of a moulding. The gel coat hides the fibre pattern of the reinforcement, protects the laminate from external influences such as ingress of moisture, and is often pigmented to provide a coloured surface.
Gel time	The time taken for a resin to set to a non-fluid gel. Sometimes called the setting time.

Hardening time	The time from the setting of the resin to the point when the resin is hard enough to allow the laminate to be withdrawn from the mould.
Maturing time	The time taken for a laminate to acquire its full hardness, chemical resistance and stability.
Monomer	A simple chemical compound capable of being converted into a polymer by combining with itself or other simple substances.
Parting agent	A lubricant which, when coated over a mould, prevents the moulding from sticking to it.
Pigment	Colouring matter, usually in the form of powder, which is added to the gel coat to impart a colour to the surface.
Polymer	A plastics material made up from large molecules formed by the joining together of simple molecules.
Post heat	Artificial heating subsequent to the start of the curing process.
Pot life	The time during which a liquid resin remains usable after curing agents have been added.
Pre-accelerated resin	A resin with accelerator pre-mixed in the correct proportions, thus requiring only the addition of catalyst. Such resins are safer to use as they prevent the inadvertent, and highly dangerous, addition of catalyst to accelerator.
Reinforcement	Usually a glass fibre membrane embedded in resin to improve the mechanical properties of the laminate.
Release agent	A parting agent. A compound which prevents the laminate from sticking to the mould and which facilitates release of the moulding.

Resin	A viscous, syrupy substance which changes to a solid when catalyst and accelerator are added.
Rovings	Continuous strands of glass wound with no appreciable twist.
Shelf life	The time during which a liquid resin can be stored under specified conditions and remain suitable for use.
Surface mat	A thin tissue of glass fibre placed on the reverse side of the moulding to improve the appearance.
Thermosetting	A material which undergoes a change (setting) when heated. Once set further heating will cause decomposition rather than softening.
Thixotropic	Firm and jelly-like at rest, but becoming fluid and mobile when stirred.
Under cure	The state of resin which has hardened unsatisfactorily, and has not attained its final chemical and physical properties.
Unsaturated resin	A resin in which the chemical reaction of curing is not complete, and which can undergo further reaction.
Viscosity	Resistance to flow. Syrup has a high viscosity, whereas water has a low viscosity.

INDEX

Other Question and Answer Boat Books

QUESTIONS AND ANSWERS STEEL BOAT CONSTRUCTION
K.A. Slade.

This book takes the reader from lofting to launching of the modern welded vessel and will appeal to students, apprentices and all shipyard workers who want to know more about ship and boat construction. The content is based on the practical experience of a man who has spent many years at one of Britain's Naval Dockyards. The sections on welding in particular reveal some invaluable information on techniques that will help the reader to achieve work of the highest standard.

1978   124 pages   165 x 111 mm   0 408 00327 8

QUESTIONS AND ANSWERS WOODEN BOAT CONSTRUCTION
R. Jurd

With its emphasis on yacht building by traditional methods this book will interest the wooden boat owner, the skilled home constructor and the craft student. Starting with the lines and construction plans the book goes on to describe step-by-step the three main forms of construction — carvel, clencher (lapstrake) and double-skin. The author describes, with the aid of sketches, the hull, the laying of decks and the construction of hatches, etc, and basic carpentry work below decks.

1978   96 pages   165 x 111 mm   0 408 00315 4

 Newnes Technical Books

Butterworths, Borough Green, Sevenoaks, Kent TN15 8PH

Question and Answer Books of Interest

CARPENTRY AND JOINERY
A.R. Whittick

1974   160 pages   165 x 111 mm   0 408 00375 8

DIESEL ENGINES
J.N. Seale

1969   128 pages   165 x 111 mm   0 600 41254 7

ELECTRIC ARC WELDING — 2nd edn
K. Leake and N.J. Henthorne

1974   132 pages   165 x 111 mm   0 408 00132 1

GAS WELDING AND CUTTING
P.H.M. Bourbousson

1973   120 pages   165 x 111 mm   0 408 00105 4

LATHEWORK
J.A. Oates

1971   128 pages   165 x 111 mm   0 408 00065 1

PIPEWORK AND PIPE WELDING
L.J. Rose

1973   108 pages   165 x·111 mm   0 408 00108 9

Butterworths, Borough Green, Sevenoaks, Kent TN15 8PH